Dedication

The author dedicate this book to scientists who contributed to the scientific fields and changed the way of our thinking.

Preface

This book created for childrens to create curiosity in science. Description of 14 scientist who made significant contributions with their picture given.

Authentic quotes and sayings of scientists are selected and included at the start of chapter. Students will find fun to read and enjoy this book

All scientist in the list are great and contributed to science in significant way.

Contents

 *Human progress has always been driven by a sense of adventure and unconventional thinking.*

Andre Konstantin Geim

Andre Konstantin Geim,born 21 October 1958 with Dutch, Russian, and British heritage. A physicist working at the University of Manchester, Geim was awarded Nobel Prize in Physics together with Konstantin Novoselov for his work on grapheme in 2010. Graphene is a super-conductive form of carbon, made from single-atom-thick sheets. Graphene consists of one-atom-thick layers of carbon atoms arranged in a two-dimensional hexagon. It is the thinnest material in the world, as well as one of the strongest and hardest and is considered a superior alternative to silicon. And has many otheruses. He also researched diamagnetic levitation and in a famous 1997 experiment, he managed to levitate a frog. He has also done research on mesoscopic physics and superconductivity. About his broad range of science that he studies, Geim says, "Many people chose a subject for their PhD and then continue the same subject until they retire. I despise this approach. I have changed my subject five times before I got my first tenured position and that helped me to learn different subjects."

Science should be fun, and you don't always need to do expensive multi-million dollar experiments to be on the cutting edge of research

Konstantin Sergeevich Novoselov

Konstantin Sergeevich Novoselov, born 23 August 1974. He is a Russian-British physicist at the University of Manchester as a Royal Society University Research Fellow. He is known for working together with Andree Geim in discovering and studying graphene. Because of their work, they won the Nobel Prize in Physics in 2010.Novoselov is also a recipient of an ERC Starting Grant from the European Research Council.

Dr. Novoselov's record includes 49 papers mostly in Physics and Materials Science and has also been cited 3,536 times in a span of twenty years. Dr. Novoselov is a Royal Society Research Fellow in School of Physics & Astronomy at the University of Manchester as well as the Langworthy Professor and director of the Manchester Centre for

Mesoscience and Nanotechnology also at the University of Manchester.

He received a Diploma from the Moscow Institute of Physics and Technology, and undertook his Ph.D. studies at the University of Nijmegen in the Netherlands before moving to the University of Manchester in the United Kingdom with his doctoral advisor Andre Geim in 2001. According to the ISI Essential Science Indicator, his two papers in Science 2004 and Nature 2005 are the most cited papers on graphene and "have opened up a fast moving front". The paper in Science 2004 is also acknowledged as "one of the most cited recent papers in the field of Physics".

Intellectual property is a key aspect for economic development.

John Craig Venter

John Craig Venter was born on 14 October 14 1946. He is an American biologist famous for being one of the first to sequence the human genome. He also created the first cell with a synthetic genome last 2010. He now works for the J. Craig Venter Institute which he founded. His current work is focused on creating synthetic biological organisms and also documenting the genetic diversity in the world's oceans. He is listed on Time magazine's 2007 and 2008 issue as part of the Time 100 list of the most influential people in the world.

The Global Ocean Sampling Expedition (GOS) is an ocean exploration genome project that will assess the genetic diversity in marine microbial life. It is to understand how the diversity contributes to nature's basic processes. The GOS circumnavigated the globe which started in 2004 and ended 2006.In May 2010, Venter and his team successfully created what they called "synthetic life". They synthesized a very long DNA molecule containing an entire bacterium genome, and introduced this into another cell. This could lead to producing bacteria that can be engineered to perform specific purpose such as create fuel, manufacture medicine, and correct environmental problems like global warming.

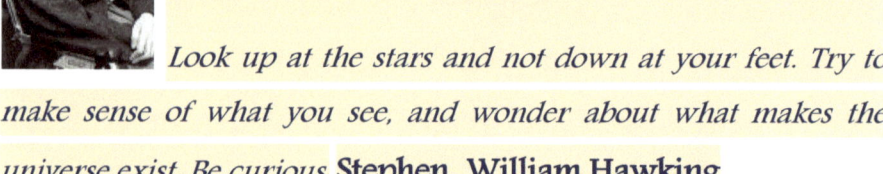 *Look up at the stars and not down at your feet. Try to make sense of what you see, and wonder about what makes the universe exist. Be curious.* **Stephen William Hawking**

Stephen William Hawking was born 8 January 1942. He is an English theoretical physicist and cosmologist. His scientific books (specially his runaway bestseller, Brief History of Time) and public appearances have made him a popular pop-icon and academic celebrity. In 2009, Hawking was awarded the Presidential Medal of Freedom, the highest civilian award in the United States.

He is known for his research and contributions to the science of cosmology and quantum gravity. He has also achieved success with works of popular science in which he discusses his own theories and cosmology in general. His contributions to science still keep coming in. Together with Roger Penrose, he provided theorems regarding gravitational singularities within the framework of general relativity. He also gave theoretical predictions about black holes emits radiation. This type of radiation is known as the Hawking radiation or the Bekenstein-Hawking radiation. Currently, he is the Director of Research at the Centre for Theoretical Cosmology in the Department of Applied Mathematics and Theoretical Physics at the University of Cambridge as well as a Fellow of Gonville and Caius College,Cambridge and a Distinguished Research Chair at the Perimeter Institute for Theoretical Physics in Waterloo, Ontario

Everything we do, every thought we've ever had, is produced by the human brain. But exactly how it operates remains one of the biggest unsolved mysteries, and it seems the more we probe its secrets, the more surprises we find. **Neil de Grasse Tyson**

Neil de Grasse Tyson was born on 5 October 1958. He is an American astrophysicist. As stated in his website, He "was born and raised in New York City where he was educated in the public schools clear through his graduation from the Bronx High School of Science. Tyson went on to earn his BA in Physics from Harvard and his PhD in Astrophysics from Columbia.".

He is the Frederick P. Rose Director of the Hayden Planetarium at the Rose Center for Earth and Space, and also a Research Associate in the Department of Astrophysics at the American Museum of Natural History. He is part of this list because of his contribution bringing astrophysics and astronomy to the public. He has hosted the educational science television show NOVA scienceNOW on PBS and has been a guest on several TV shows such as The Daily Show, The Colbert Report, and Jeopardy!. It was announced that Tyson will be hosting a new sequel to Carl Sagan's Cosmos: A Personal Voyage TV series. He has made astronomy an interesting subject to people worldwide.

One in 200 stars has habitable Earth-like planets surrounding it-in the galaxy, half a billion stars have Earth-like planets going around them – that's huge, half a billion. So when we look at the night sky, it makes sense that someone is looking back at us.

Michio Kaku

Michio Kaku was born on 24 January 1947. He is an American theoretical physicist. A Henry Semat Professor of Theoretical Physics in the City College of New York of City University of New York, he is also the co-founder of string field theory. Another science communicator like Neil DeGrasse Tyson, Michio Kaku has written several books about physics and related topics.He has also made guest appearances on radio, television, and film.

Kaku achieved popularity because of his knowledge and easy approach to explaining complicated science subjects such as time travel and singularities. Although a theoretical physicist, he covers a wide range of subjects such as wormholes and time travel. He considered the theory that the universe was created from nothing as discussed in the TV show, "What Happened Before the Big Bang". There are more scientists out there who have contributed to our society. These are just a few of them. We're still at the dawn of this new century. But our science have moved leaps and strides faster than the previous one.

In the process of evolution, the body lasts for some time and then will take other body and take other body and take other body until the final redemption from diversity is transcended.

Maharishi Mahesh Yogi

Maharishi Mahesh Yogi (12 January 1918[1] – 5 February 2008) was born Mahesh Prasad Varma and became known as Maharishi (meaning "great seer") and Yogi as an adult. He developed the Transcendental Meditation technique and was the leader and guru of a worldwide organization that has been characterized in multiple ways including as a new religious movement and as non-religious. The Maharishi is credited as having contributed to the western world a meditation technique that is both simple and systematic as well as introducing the scientific study of meditation. Maharishi stated that the experience of transcendence, which resulted in a naturally increasing refinement of mind and body, enabled people to naturally behave in more correct ways .Thus, behavioral guidelines did not need to be issued, and were best left to the teachings of various religions: "It is much easier to raise a man's consciousness than to get him to act righteously" the Maharishi said. In his 1967 publication, Bhagavad-Gita:A New Translation and Commentary, the Maharishi describes the Bhagavad Gita as "the Scripture of Yoga". He says that "its purpose is to explain in theory and practice all that is needed to raise the consciousness of man to the highest possible level. According to Peter Russell, the Bhagavad-Gita deals with the concept of loss of knowledge and subsequent revival, and this is brought out by the Maharishi himself in the introduction. In the Preface, the Maharishi writes: "The purpose of this commentary is to restore the fundamental truths of the Bhagavad-Gita and thus restore the significance of its teaching. If this teaching is followed, effectiveness in life will be achieved, men will be fulfilled on all levels and the historical need of the age will be fulfilled also.

Governments and scientists in India need to ensure that politics and religious ideology do not intrude into science. They belong to separate spheres, and if they are not kept separate, it is science in India and the country as a whole that will suffer

Venkatraman "Venki" Ramakrishnan

Venkatraman "Venki" Ramakrishnan (born 1952) is an Indian-American-British structural biologist of Indian origin. He is the current President of the Royal Society, having held the position since November 2015. In 2009 he shared the Nobel Prize in Chemistry with Thomas A. Steitz and Ada Yonath, "for studies of the structure and function of the ribosome". Since 1999, he has worked as a group leader at the Medical Research Council (MRC) Laboratory of Molecular Biology (LMB) on the Cambridge Biomedical Campus, UK, where he is also the Deputy Director Ramakrishnan was born in Chidambaram in Cuddalore district of Tamil Nadu, India to C.V.Ramakrishnan and Rajalakshmi Ramakrishnan in a Tamil Iyer family. Both his parents were scientists, and his father was head of department of biochemistry at the Maharaja Sayajirao University of Baroda. At the time of his birth, Ramakrishnan's father was away from India doing postdoctoralresearch with David E. Green at the University of Wisconsin–Madison.

His mother obtained a PhD in Psychology from McGill University in 1959 which she completed in only 18 months, and was mentored by Donald O.Hebb.LalitaRamakrishnan, his younger sister, is professor of immunology and infectious diseases at the Department of Medicine, University of Cambridge, and is a member of the National Academy of Sciences.

Ramakrishnan began work on ribosomes as a postdoctoral fellow with Peter Moore at Yale University.After his post-doctoral fellowship, he initially could not find a faculty position even though he had applied to about 50 universities in the U.S.

He continued to work on ribosomes from 1983-95 as a staff scientist at Brookhaven National Laboratory. In 1995 he moved to the University of Utah as a Professor of Biochemistry , and in 1999, he moved to his current position at the Medical Research Council Laboratory of Molecular Biology in Cambridge, England, where he had also been a sabbatical visitor during 1991-92

Half a century ago, he was worrying about climate change. He was a pioneer in electric car technology

Stanford Robert Ovshinsky

Stanford Robert Ovshinsky, a prolific American inventor and scientist who had a special passion for renewable energy, passed away in 2012 .Mr. Ovshinsky had over 400 patents to his name, including nickel metal hydride (NiMH) batteries – still used in hybrid electric cars today – continuous web multi-junction flexible thin-film solar laminates and hydrogen fuel cells.Ovshinsky created a hatful of world-changing innovations. In 1968, the New York Times declared that his new electronic switch would lead to a future in which we would all have "small, general-purpose desktop computers for use in homes, schools and offices" and "a flat, tubeless television set that can be hung on the wall like a picture".

It seemed so unlikely that no one in the US wanted to invest. What's more, Ovshinsky's discoveries threatened the dominance of America's great new invention: the transistor. US corporate interests rubbished his work and he ended up licensing his technologies to a few small Japanese companies.

False facts are highly injurious to the progress of science, for they often endure long; but false views, if supported by some evidence, do little harm, for everyone takes a salutary pleasure in proving their falseness.

Charles Robert Darwin,

12 February 1809 – 19 April 1882) was an English naturalist and geologist, best known for his contributions to the science of evolution He established that all species of life have descended over time from common ancestors, and in a joint publication with Alfred Russel Wallace introduced his scientific theory that this branching pattern of evolution resulted from a process that he called natural selection, in which the struggle for existence has a similar effect to the artificial selection involved in selective breeding.

Darwin published his theory of evolution with compelling evidence in his 1859 book On the Origin of Species, overcoming scientific rejection of earlier concepts of transmutation of species. By the 1870s, the scientific community and much of the general public had accepted evolution as a fact. However, many favoured competing explanations and it was not until the emergence of the modern evolutionary synthesis from the 1930s to the 1950s that a broad consensus developed in which natural selection was the basic mechanism of evolution. In modified form, Darwin's scientific discovery is the unifying theory of the life sciences, explaining the diversity of life.

I decided that my means were sufficient to enable me to devote myself to botany, a determination which I never, during the long period of my subsequent career, had on any occasion any reason to repent of. **George Bentham**

Bentham was born in Stoke, Plymouth, on 22 September 1800. His father, Sir Samuel Bentham, a naval architect, was the only brother of Jeremy Bentham to survive into adulthood. George Bentham had neither a school nor a college education, but at an early age acquired the power of giving sustained and concentrated attention to any subject that occupied him. He also had a remarkable linguistic aptitude. By the age of seven he could speak French, German and Russian, and he learned Swedish during a short residence in Sweden when little older. At the close of the war with France, the Benthams made a long tour through that country, staying two years at Montauban, where Bentham studied Hebrew and mathematics in the Protestant Theological School. They eventually settled in the neighbourhood of Montpellier where Sir Samuel purchased a large estate.

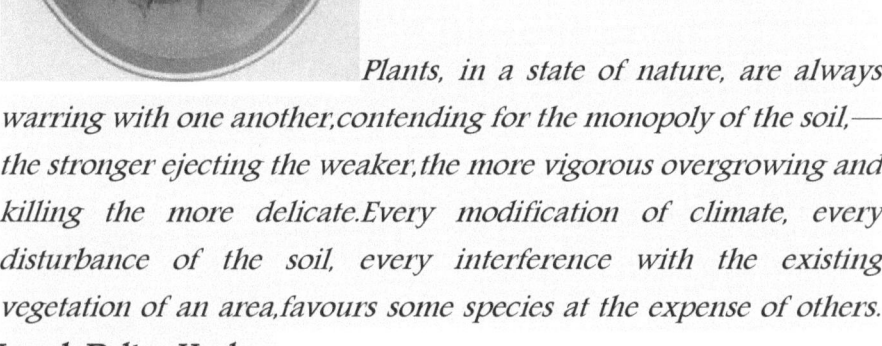

Plants, in a state of nature, are always warring with one another,contending for the monopoly of the soil,— the stronger ejecting the weaker,the more vigorous overgrowing and killing the more delicate.Every modification of climate, every disturbance of the soil, every interference with the existing vegetation of an area,favours some species at the expense of others.

Joseph Dalton Hooker

Sir Joseph Dalton Hooker (1817-1911), botanist and explorer, was born on 30 June 1817 at Halesworth, Suffolk, England, second son of the distinguished botanist, Sir William Jackson Hooker (1785-1865), and his wife Maria Sarah, eldest daughter of Dawson Turner, banker and naturalist of Norwich. His father, later director of the Royal Botanic Gardens, Kew, was, from 1820 to 1841 Regius professor of botany in the University of Glasgow and Joseph was educated at Glasgow Grammar School. At 15 he began to attend classes at the University of Glasgow, at first in classics and mathematics and later in medicine (M.D., 1839).

He already had a wide knowledge of botany based on work in his father's herbarium and on extensive plant-collecting in the British Isles. His degree enabled him to join the Naval Medical Service and to accompany a scientific expedition to the Antarctic. The expedition, commanded by James Clark Ross, sailed in 1839 in two ships, Erebus and Terror: Hooker was assistant surgeon and naturalist in the former. They visited Ascension, St Helena, the Cape, Kerguelen, Van Diemen's Land, New Zealand, Tierra del Fuego and the Falkland Islands, and sailed along a vast extent of the coast of Antarctica. Van Diemen's Land was visited twice, during August–October 1840 and March–May 1841, and there was a brief stay at Port Jackson. The expedition returned to England in 1843.

Our greatest weakness lies in giving up.The most certain way to succeed is always to try just one more time.

Thomas Alva Edison

Thomas Alva Edison (February 11, 1847 – October 18, 1931) was an American inventor and businessman. He developed many devices that greatly influenced life around the world, including the phonograph, the motion picture camera, and the long-lasting, practical electric light bulb. Dubbed "The Wizard of Menlo Park", he was one of the first inventors to apply the principles of mass production and large-scale teamwork to the process of invention, and because of that, he is often credited with the creation of the first industrial research laboratory.

*When we come close to those things that break us down, we touch those things that also break us open. And in that breaking open, we uncover our true nature.*

Richard A. Muller

Richard A. Muller (born January 6, 1944) is an American professor of physics at the University of California, Berkeley. He is also a faculty senior scientist at the Lawrence Berkeley National Laboratory. Most recently, in early 2010, Muller and his daughter Elizabeth founded the group Berkeley Earth, an independent 501(c)(3) non-profit aimed at addressing some of the major concerns of the climate change sceptics, in particular the global surface temperature record.

www.ingramcontent.com/pod-product-compliance
Lightning Source LLC
Chambersburg PA
CBHW041615180526
45159CB00002BC/871